Quantum

Compute

Volume II

Unleashing the Power of
the Future

Dr. Robert Safadi-Ph.D.

Table Of Contents

Chapter 1: Introduction to Quantum Computing

The Need for Quantum Computing

In today's digitally-driven world, the need for advanced computing technologies has never been more critical. As we continue to push the boundaries of scientific research, artificial intelligence, cryptography, and data analysis, traditional computers are reaching their limits in terms of processing power. This is where quantum computing comes in – a revolutionary technology that has the potential to unlock unimaginable possibilities and transform the way we approach and solve complex problems.

Quantum computing harnesses the principles of quantum mechanics to process information in a fundamentally different way than classical computers.

While classical computers use binary bits (0s and 1s) to store and manipulate data, quantum computers utilize quantum bits, or qubits, which can exist in multiple states simultaneously thanks to a phenomenon called superposition.

The Quantum Computing Revolution

This unique property allows quantum computers to perform intricate calculations and solve problems that are currently beyond the reach of classical computers.

One of the most promising applications of quantum computing lies in the field of cryptography. As our reliance on digital communication and data storage increases, so does the need for robust encryption methods to protect sensitive information. Quantum computers have the potential to crack existing encryption algorithms that are used to secure our data, rendering them vulnerable to cyber-attacks. However, they can also provide us with new, unbreakable encryption methods that leverage the principles of quantum mechanics. This makes quantum computing crucial for maintaining privacy and security in the digital age.

Another area where quantum computing holds immense promise is in the optimization of complex systems.

Many real-world problems, such as route optimization, supply chain management, and financial portfolio optimization, involve countless variables and constraints that make finding the best solution a daunting task for classical computers.

The Quantum Computing Revolution

Quantum computers, with their ability to process and analyze vast amounts of information simultaneously, could revolutionize these fields by providing optimal solutions in a fraction of the time it would take classical computers.

Furthermore, quantum computing has the potential to revolutionize scientific research by simulating complex phenomena that are currently impossible to model accurately.

From understanding the behavior of molecules for drug discovery to unraveling the mysteries of the universe, quantum computers could enable us to simulate and explore the intricacies of nature in unprecedented ways.

In conclusion, the need for quantum computing is evident across various fields and industries.

Its ability to solve complex problems, enhance encryption, optimize systems, and revolutionize scientific research makes it a technology worth investing in. As we unlock the power of quantum computing, we will pave the way for a future where unimaginable advancements become a reality.

The Quantum Computing Revolution

Whether you are a quantum computing enthusiast or simply curious about the potential of this groundbreaking technology, exploring the world of quantum computing will undoubtedly leave you in awe of the limitless possibilities it holds.

History and Evolution of Quantum Computing

In the rapidly advancing world of technology, quantum computing stands as one of the most groundbreaking and promising fields to date.

With the potential to revolutionize everything from cryptography to drug discovery, it is no wonder that quantum computing has captured the attention of scientists, researchers, and tech enthusiasts worldwide.

This subchapter aims to provide a comprehensive overview of the fascinating history and evolution of quantum computing, shedding light on its origins and the significant milestones that have paved the way for its current state.

The Quantum Computing Revolution

The journey of quantum computing began in the early 20th century with the advent of quantum mechanics. Pioneers such as Max Planck, Albert Einstein, and Erwin Schrödinger laid the foundation for the field, introducing the concept of quantum superposition and the existence of qubits, the fundamental unit of quantum information. However, it wasn't until the 1980s that the idea of harnessing quantum phenomena for computing purposes gained traction.

In 1981, Nobel laureate Richard Feynman proposed the concept of a quantum computer, highlighting its potential to simulate quantum systems more efficiently than classical computers.

This sparked a wave of research in the field, leading to several breakthroughs in the 1990s. Notably, Peter Shor's discovery of a quantum algorithm capable of factoring large numbers exponentially faster than classical algorithms demonstrated the immense computational power quantum computers could offer.

The Quantum Computing Revolution

The early 2000s witnessed significant progress in the development of quantum computing hardware. Researchers started experimenting with various technologies, including superconducting circuits, trapped ions, and topological qubits, each offering unique advantages and challenges. Companies like IBM, Google, and Microsoft joined the race, investing heavily in quantum research and development.

Today, quantum computers are becoming more accessible, thanks to cloud-based platforms and companies offering quantum computing as a service.

These advances have allowed researchers and developers from diverse fields to explore the potential applications of quantum computing, ranging from optimization problems and machine learning to drug discovery and material science.

While quantum computing is still in its infancy, the future holds immense promise. As scientists continue to refine quantum algorithms, improve qubit stability, and overcome various technical challenges, we are inching closer to a quantum revolution.

The Quantum Computing Revolution

This subchapter aims to provide readers with a comprehensive understanding of the history and evolution of quantum computing, enabling them to appreciate the significance of this revolutionary technology and its potential to transform industries and society as a whole.

Whether you are a quantum computing enthusiast, a tech-savvy individual, or simply curious about the future of computing, this subchapter will serve as an enlightening guide, equipping you with the knowledge to comprehend the remarkable journey of quantum computing and its immense potential for the future.

Basic Principles of Quantum Mechanics

Quantum mechanics is the fundamental theory that underlies the field of quantum computing. It is a branch of physics that delves into the behavior and properties of particles at the quantum level. Understanding the basic principles of quantum mechanics is crucial for anyone interested in quantum computing, as it forms the foundation upon which this revolutionary technology is built.

At its core, quantum mechanics challenges our classical understanding of the physical world. Unlike classical physics, which describes objects as having definite properties and behaviors, quantum mechanics introduces the concept of superposition.

Superposition allows particles to exist in multiple states simultaneously, making quantum systems inherently different from their classical counterparts.

Another key principle of quantum mechanics is entanglement.

The Quantum Computing Revolution

Entanglement occurs when two or more particles become connected in such a way that the state of one particle is dependent on the state of the other(s). This phenomenon is often referred to as "spooky action at a distance" and is one of the most intriguing aspects of quantum mechanics.

In addition to superposition and entanglement, quantum mechanics also introduces the notion of wave-particle duality. This principle suggests that particles, such as electrons or photons, can exhibit both wave-like and particle-like behaviors, depending on the experimental setup. This duality challenges our classical intuition and highlights the unique nature of quantum systems.

Quantum mechanics is governed by mathematical equations known as wave functions, which describe the probabilities of different outcomes when measuring quantum particles. These wave functions are typically represented by complex numbers and can be used to calculate properties such as energy levels and probabilities of measurements.

While the principles of quantum mechanics may initially appear abstract and counterintuitive, they provide the basis for the immense power of quantum computing. By harnessing the principles of superposition and entanglement, quantum computers have the potential to solve complex problems exponentially faster than classical computers.

In conclusion, understanding the basic principles of quantum mechanics is essential for anyone interested in quantum computing. Superposition, entanglement, and wave-particle duality are the key concepts that differentiate quantum systems from classical ones.

By grasping these principles, individuals can appreciate the immense potential of quantum computing and its ability to revolutionize various fields, including cryptography, optimization, and drug discovery. The journey into the world of quantum computing starts with a solid foundation in the basic principles of quantum mechanics.

Chapter 2: Fundamentals of Quantum Computing

Quantum Bits (Qubits)

Quantum Bits (Qubits): Unleashing the Power of Quantum Computing

In the ever-evolving landscape of technology, quantum computing has emerged as a promising field that holds immense potential to revolutionize the way we process information. At the heart of quantum computing lies the concept of Quantum Bits or Qubits, which form the building blocks of this revolutionary technology.

Qubits are the fundamental units of information in quantum computing, analogous to classical bits in traditional computing.

However, unlike classical bits that can only represent a value of 0 or 1, qubits can exist in a superposition of both states simultaneously. This unique property of qubits allows quantum computers to perform computations exponentially faster than their classical counterparts.

The Quantum Computing Revolution

To understand the power of qubits, let's dive deeper into their underlying principles. Qubits can be implemented using various physical systems, such as atoms, ions, superconducting circuits, or even photons. These systems harness the principles of quantum mechanics to manipulate and store information.

One of the most remarkable aspects of qubits is entanglement. Entanglement is a phenomenon where two or more qubits become interconnected, creating an inseparable relationship.

When qubits are entangled, the state of one qubit directly affects the state of the other, regardless of the distance between them. This property allows quantum computers to perform parallel computations, exponentially increasing their computational power.

Quantum computing's potential applications are vast and diverse.

From optimizing complex logistics and financial systems to revolutionizing drug discovery and solving complex mathematical problems, quantum computers can tackle challenges that are currently beyond the reach of classical computers.

The Quantum Computing Revolution

However, harnessing the power of qubits is not without its challenges. Qubits are highly sensitive to environmental disturbances, such as noise and temperature fluctuations, which can cause errors in computations. Overcoming these challenges requires the development of robust error correction techniques and efficient qubit designs.

In conclusion, qubits are the key to unlocking the true potential of quantum computing. Their ability to exist in superposition and be entangled provides a pathway toward solving complex problems exponentially faster than classical computers.

As quantum computing continues to advance, we can expect to witness a transformative revolution in various fields, from finance to healthcare and beyond. The era of quantum computing is upon us, and the power of qubits will shape the future of computing.

Quantum Gates and Operations

In the fascinating realm of quantum computing, one of the fundamental concepts that form the backbone of its power and potential is the notion of quantum gates and operations.

These concepts play a crucial role in manipulating and transforming the quantum bits, or qubits, which are the building blocks of quantum computers. Understanding quantum gates and operations is essential for anyone interested in delving deeper into the world of quantum computing.

At its core, a quantum gate is a mathematical operation that acts on one or more qubits, altering their quantum states. Just like classical logic gates, which perform basic operations on classical bits to carry out computations, quantum gates serve as the basic building blocks for quantum computations. However, quantum gates possess unique properties due to the principles of superposition and entanglement that govern the behavior of qubits.

The Quantum Computing Revolution

One of the most well-known and fundamental quantum gates is the Hadamard gate. When applied to a qubit, the Hadamard gate places it in a superposition of both 0 and 1 states. This superposition allows quantum computers to process vast amounts of information simultaneously, providing them with the potential for exponential speedup over classical computers.

Another crucial quantum gate is the CNOT gate or the controlled-NOT gate. This gate introduces entanglement between two qubits, where the target qubit's state is flipped only if the control qubit is in state 1. The ability to entangle qubits is a defining feature of quantum computing, enabling powerful operations like quantum teleportation and quantum error correction.

Other essential quantum gates include the Pauli-X, Pauli-Y, and Pauli-Z gates, which rotate qubit states around different axes in the Bloch sphere. These gates allow for the manipulation of quantum information in various ways, forming the basis for more complex quantum operations.

The Quantum Computing Revolution

Understanding quantum gates and operations is crucial for harnessing the power of quantum computing. By combining different quantum gates, one can design and execute complex quantum algorithms that outperform classical counterparts in specific tasks, such as factoring large numbers or simulating quantum systems.

As the field of quantum computing continues to evolve, researchers and engineers are exploring new types of quantum gates and operations, pushing the boundaries of what is possible. These advancements promise to unlock the true potential of quantum computing, revolutionizing industries such as cryptography, drug discovery, optimization, and more.

Whether you are a quantum computing enthusiast, a researcher, or simply curious about the future of computing, delving into the world of quantum gates and operations is an essential step in understanding the power and transformative capabilities of this revolutionary technology.

Quantum Circuits and Algorithms

In this subchapter, we will explore the fascinating world of quantum circuits and algorithms, unveiling the potential they hold in revolutionizing the field of computing. Whether you are a seasoned quantum computing enthusiast or just starting your journey into this exciting domain, this chapter will provide you with a comprehensive understanding of the principles behind quantum circuits and algorithms.

Quantum circuits are the building blocks of quantum computers, analogous to classical computing's logic gates. However, they operate on quantum bits, or qubits, which can exist in multiple states simultaneously due to the phenomenon of superposition. Harnessing this superposition and another quantum property called entanglement enables quantum computers to perform calculations at an exponentially higher speed than classical computers.

We will delve into the fundamental components of a quantum circuit, such as quantum gates, which manipulate the qubits' states.

The Quantum Computing Revolution

These gates include the Hadamard gate, CNOT gate, and many others, each serving a specific purpose in quantum computations. Understanding these gates is crucial for designing and implementing quantum algorithms.

Speaking of algorithms, we will explore some of the most influential quantum algorithms developed to date.

One of the most famous examples is Shor's algorithm, which demonstrates quantum computing's ability to solve certain mathematical problems exponentially faster than classical computers. This algorithm has the potential to break modern encryption schemes, making it a subject of great interest and concern.

We will also discuss Grover's algorithm, a powerful quantum search algorithm that can efficiently solve unstructured search problems. This algorithm has implications in various fields, including optimization, data mining, and artificial intelligence.

Throughout this subchapter, we will provide intuitive explanations and concrete examples to ensure that readers from all backgrounds can grasp the underlying concepts. Moreover, we will discuss the current challenges faced in implementing quantum circuits and algorithms, such as noise, decoherence, and error correction techniques.

By the end of this subchapter, you will have a solid foundation in quantum circuits and algorithms, enabling you to appreciate the potential of quantum computing in various fields.

Whether you are a quantum computing enthusiast, a researcher, or simply curious about the future of computing, this chapter will unlock the power of the quantum computing revolution for you. Get ready to embark on an extraordinary journey into the realm of quantum computation!

Chapter 3: Quantum Computing Hardware

Quantum Processors and Quantum Registers

In this subchapter, we will delve into the fascinating world of quantum processors and quantum registers, two key components of quantum computing that are driving the future of technology. Whether you are a quantum computing enthusiast or someone who is just beginning to understand this revolutionary field, this content will provide valuable insights into the inner workings of quantum processors and registers.

Quantum processors lie at the heart of quantum computers, acting as the engines that execute complex algorithms at incredible speeds. Unlike classical computers that rely on bits to store and process information, quantum processors leverage the power of quantum bits or qubits. These qubits can exist in multiple states simultaneously, thanks to a property called superposition. Harnessing this superposition, quantum processors can perform parallel computations, exponentially increasing computational power.

The Quantum Computing Revolution

However, the delicate nature of qubits poses significant challenges. Quantum processors must be shielded from external disturbances, as even the slightest interference can cause decoherence and lead to errors in calculations. Researchers are constantly working to develop error correction techniques to mitigate these issues, as they are crucial for the scalability and reliability of quantum computers.

To operate quantum processors effectively, quantum registers are employed. Quantum registers act as the memory of quantum computers, holding the qubits in a stable state until they are needed for computation. Similar to classical registers, quantum registers allow for the storage and retrieval of information. However, unlike classical bits, qubits in quantum registers can be entangled, enabling the creation of highly complex states that are crucial for quantum algorithms.

Quantum registers come in various forms, such as superconducting circuits, trapped ions, or topological qubits. Each type has its advantages and disadvantages, and scientists are continually exploring new avenues to create more stable and efficient quantum registers.

The Quantum Computing Revolution

Understanding the intricacies of quantum processors and registers is vital for anyone interested in the field of quantum computing.

As the demand for faster and more powerful computers continues to grow, quantum processors hold the promise of revolutionizing industries like cryptography, drug discovery, and optimization problems.

By harnessing the power of quantum bits and quantum registers, we can unlock the limitless potential of quantum computing and pave the way for a new era of technological advancement.

Whether you are a quantum computing enthusiast, a researcher, or simply curious about the future of computing, exploring the possibilities offered by quantum processors and registers will undoubtedly expand your understanding of this groundbreaking field and inspire you to be a part of the quantum computing revolution.

Types of Quantum Computers

In the rapidly evolving field of quantum computing, there are several different types of quantum computers that researchers and engineers are exploring. Each type has its own unique characteristics and potential applications, bringing us closer to unlocking the power of the future. In this subchapter, we will delve into some of the most promising types of quantum computers and their key features.

1. Quantum Annealers:
Quantum annealers are one of the earliest types of quantum computers to be developed. They excel at solving optimization problems and are particularly well-suited for tasks like simulating physical systems, financial modeling, and machine learning.

These computers leverage a process called quantum annealing to find the best solution among a vast number of possibilities.

2. Gate-Based Quantum Computers:
Gate-based quantum computers are often referred to as universal quantum computers because they can perform any computation that a classical computer can, given enough qubits and resources.

The Quantum Computing Revolution

These computers rely on quantum gates to manipulate qubits, which can be thought of as the building blocks of quantum information. Gate-based quantum computers hold immense potential for solving complex problems in cryptography, chemistry, and materials science.

3. Quantum Simulators:
Quantum simulators are a specialized type of quantum computer that aim to simulate and study complex quantum systems. By accurately modeling and exploring the behavior of quantum systems, researchers can gain insights into phenomena that are difficult to observe through classical means. Quantum simulators have the potential to revolutionize fields such as quantum chemistry, condensed matter physics, and quantum biology.

4. Topological Quantum Computers:
Topological quantum computers are based on a branch of physics called topological quantum field theory. They rely on exotic particles called anyons, which can be manipulated to perform quantum computations.

The Quantum Computing Revolution

Topological quantum computers are highly robust against errors and noise, making them a promising avenue for fault-tolerant quantum computing. However, they are still in the early stages of development, and many technical challenges need to be overcome.

5. Quantum Neural Networks:
Quantum neural networks combine the power of quantum computing with artificial intelligence techniques. Inspired by classical neural networks, these quantum counterparts have the potential to revolutionize machine learning and pattern recognition. Quantum neural networks can process vast amounts of data simultaneously and may lead to breakthroughs in areas such as natural language processing, image recognition, and data analysis.

As the field of quantum computing continues to advance, these different types of quantum computers will play crucial roles in various applications. While each type has its own strengths and limitations, their combined potential promises to reshape the future of computing, revolutionizing industries and solving complex problems that are currently beyond the reach of classical computers.

Superconducting Quantum Computing

In the ever-evolving landscape of technology, quantum computing stands as the epitome of innovation and promises to revolutionize the way we process information. Among the various approaches to quantum computing, one that has gained significant traction is Superconducting Quantum Computing. This subchapter aims to demystify the concept and shed light on its potential applications and implications for the future.

Superconducting Quantum Computing leverages the mind-boggling principles of quantum mechanics to perform computations at an unprecedented speed.

Unlike classical computers that rely on binary bits, quantum computers employ quantum bits or qubits, which can exist in multiple states simultaneously, thanks to a phenomenon called superposition. Superposition allows qubits to represent and process vast amounts of data simultaneously, leading to exponential computational power.

The Quantum Computing Revolution

One of the main advantages of superconducting qubits is their compatibility with existing semiconductor fabrication techniques. This compatibility enables the integration of superconducting qubits with classical computer components, paving the way for practical implementation and scalability.
Several tech giants and research institutions have invested heavily in developing superconducting quantum computers, making significant progress toward building large-scale, fault-tolerant quantum systems.

The potential applications of superconducting quantum computing span numerous fields.

 From optimizing complex logistical operations to simulating molecular interactions for drug discovery, superconducting quantum computers have the potential to revolutionize industries and solve problems that are currently intractable for classical computers.

Additionally, the encryption-breaking capabilities of quantum computers pose both a threat and an opportunity for cybersecurity, prompting researchers to explore post-quantum cryptography methods.

The Quantum Computing Revolution

While superconducting quantum computing holds immense promise, it also faces challenges.

Qubits are incredibly fragile, susceptible to noise, and prone to errors caused by decoherence. Researchers are actively working on developing error-correction techniques and quantum error-correcting codes to mitigate these issues and improve the stability and reliability of qubits.

In conclusion, Superconducting Quantum Computing represents a significant leap forward in the realm of quantum computing. Its compatibility with existing technology and potential for solving complex problems make it an exciting prospect for the future. As technology continues to advance, the possibilities for quantum computing are boundless.

Whether you are a quantum computing enthusiast or someone curious about cutting-edge developments in technology, exploring the world of superconducting quantum computing will undoubtedly ignite your imagination and showcase the limitless potential of the quantum computing revolution.

Topological Quantum Computing

In recent years, the field of quantum computing has gained significant attention due to its potential to revolutionize various industries and scientific fields.

One promising approach within the realm of quantum computing is Topological Quantum Computing (TQC), which offers unique advantages and groundbreaking possibilities for solving complex computational problems.

TQC is based on the concept of topological states of matter, which exhibit robust and error-resistant properties. Unlike classical computing, which relies on binary bits, quantum computing utilizes quantum bits or qubits.

These qubits can exist in multiple states simultaneously, thanks to a phenomenon called superposition. This property allows quantum computers to perform calculations at an exponentially faster rate than classical computers.

The Quantum Computing Revolution

One of the key challenges in quantum computing is maintaining the fragile quantum states while minimizing errors caused by external disturbances. TQC addresses this challenge by leveraging topological properties, which are unaffected by local perturbations. This inherent resilience to errors makes TQC a promising avenue for the development of fault-tolerant quantum computers.

The core building blocks of TQC are anyons, which are exotic particles that emerge in certain topological systems. These anyons possess unique properties such as non-Abelian statistics, where the order in which they are manipulated affects the outcome of the computation.

By braiding or manipulating these anyons, quantum gates can be implemented, enabling the encoding and manipulation of quantum information.

One of the most intriguing aspects of TQC is the potential for creating topological qubits, also known as Majorana zero modes. These qubits are highly stable and immune to environmental noise, making them ideal for large-scale quantum computing systems.

The Quantum Computing Revolution

Furthermore, TQC has the potential to address one of the biggest challenges in quantum computing: quantum error correction. By utilizing the topological properties of anyons, TQC can provide a framework for error-free quantum computation by detecting and correcting errors that occur during the computation.

While TQC is still in its early stages of development, it holds immense promise for enabling more robust and efficient quantum computing systems. As researchers continue to explore the potential applications of TQC, it is becoming increasingly clear that topological quantum computers could unlock the power of the future and revolutionize various fields, including cryptography, optimization, and drug discovery.

In conclusion, Topological Quantum Computing represents a groundbreaking approach within the field of quantum computing. By leveraging topological properties, TQC offers the potential for error-resistant quantum computation and the creation of stable topological qubits.

As the research progresses, TQC may pave the way for practical quantum computers capable of tackling complex computational problems that are currently beyond the reach of classical computers.

Chapter 4: Quantum Computing Applications

Optimization and Simulation

In the rapidly evolving world of quantum computing, one of the most promising applications is optimization and simulation.

This subchapter explores how this groundbreaking technology is revolutionizing the fields of optimization and simulation, offering tremendous potential for industries and researchers alike.

Whether you are a quantum computing enthusiast or a professional in a specific niche, this section will provide you with valuable insights into the power of quantum computing.

Optimization is a fundamental problem-solving approach in various domains, from finance and logistics to healthcare and manufacturing.

The Quantum Computing Revolution

Traditional computing methods often struggle when faced with complex optimization problems, as they require extensive computational resources and may take an impractical amount of time to reach a solution. However, quantum computing brings a ray of hope by leveraging quantum algorithms such as the Quantum Approximate Optimization Algorithm (QAOA) and the Quantum Alternating Operator Ansatz (QAOA). These algorithms harness the principles of quantum mechanics to explore multiple solutions simultaneously, drastically reducing the time required to find the optimal solution.

Simulations, on the other hand, play a crucial role in scientific research, engineering, and even entertainment. From predicting weather patterns and simulating molecule behavior to designing efficient energy systems, simulations help us understand and predict complex phenomena.

Quantum computers have the potential to revolutionize simulations by simulating quantum systems themselves. As quantum systems are inherently quantum mechanical, classical computers often struggle to accurately model their behavior.

The Quantum Computing Revolution

Quantum simulators, however, can simulate quantum systems with unparalleled precision, opening new frontiers in scientific discovery and technological advancements.

The optimization and simulation capabilities of quantum computing have the potential to transform various industries. Financial institutions can harness quantum computing to optimize investment portfolios, minimize risks, and improve trading strategies. In healthcare, quantum optimization algorithms can enhance drug discovery processes, enabling faster and more effective treatments.

Furthermore, quantum simulation can revolutionize materials science, allowing researchers to design new materials with extraordinary properties, such as superconductivity or high-strength alloys.

In conclusion, optimization and simulation are two critical areas where quantum computing is poised to revolutionize problem-solving. Its ability to explore multiple solutions simultaneously and simulate quantum systems accurately offers unprecedented potential across industries and scientific research.

The Quantum Computing Revolution

\Whether you are a quantum computing enthusiast or a professional in a specific niche, understanding the power of optimization and simulation in quantum computing is key to unlocking the immense possibilities of this transformative technology.

Cryptography and Data Security

In today's digital age, where our lives are increasingly intertwined with technology, the need for robust data security has never been more critical.

From financial transactions to personal information, our data is constantly at risk of being compromised.

This subchapter on "Cryptography and Data Security" aims to explore the fundamental principles of data protection and introduce the role of quantum computing in revolutionizing this field.

Cryptography, the practice of secure communication, has been around for centuries, evolving from ancient methods like Caesar ciphers to complex algorithms used today.

The Quantum Computing Revolution

It involves encoding information in such a way that it becomes unintelligible to unauthorized users, ensuring confidentiality and integrity.

This subchapter will delve into the basics of symmetric and asymmetric encryption, public and private keys, digital signatures, and the importance of key management.

However, as technology advances, so do the methods employed by malicious actors to breach data security.

Quantum computing, a disruptive technology on the horizon, has the potential to render many existing cryptographic techniques useless. Quantum computers harness the power of quantum mechanics to perform computations at an unprecedented speed, making it possible to crack encryption algorithms that would take classical computers years or even centuries to decipher.

The emergence of quantum computing has necessitated the development of quantum-resistant cryptographic algorithms, known as post-quantum cryptography (PQC).

The Quantum Computing Revolution

This subchapter will shed light on the ongoing efforts to create PQC standards that can withstand attacks from both classical and quantum computers. It will explore lattice-based, code-based, and multivariate-based encryption schemes, among others, highlighting their resilience against quantum attacks.

Furthermore, this subchapter will discuss how quantum computing can also be leveraged to enhance data security.

Quantum key distribution (QKD), for instance, enables the distribution of cryptographic keys with provable security based on the laws of quantum physics. Quantum random number generators (QRNGs) can provide true randomness, a crucial element in encryption algorithms.

These quantum-enabled solutions have the potential to revolutionize data security, ensuring that our sensitive information remains protected in the quantum era.

In conclusion, "Cryptography and Data Security" is a subchapter that provides a comprehensive overview of the principles and practices of data protection.

It introduces the audience to the basic concepts of encryption and the challenges posed by quantum computing.

By exploring the world of post-quantum cryptography and quantum-enabled solutions, this subchapter aims to equip individuals in the Quantum Compute niche with the knowledge required to navigate the evolving landscape of data security in the quantum computing revolution.

Machine Learning and Artificial Intelligence

In the fast-paced world of technology, two terms that have gained significant attention are Machine Learning (ML) and Artificial Intelligence (AI).

These transformative technologies have the potential to revolutionize various industries, including quantum computing.

This subchapter aims to provide a comprehensive understanding of ML and AI and their implications in the quantum computing realm.

The Quantum Computing Revolution

Machine Learning refers to the ability of computer systems to learn and improve from experience without being explicitly programmed.

It involves the development of algorithms and models that allow computers to analyze vast amounts of data, identify patterns, and make informed decisions. ML has already demonstrated its potential in various fields, such as image recognition, natural language processing, and data analytics.

Artificial Intelligence, on the other hand, encompasses broader concepts that involve the development of intelligent machines capable of simulating human intelligence.

AI systems not only learn from data but also possess the ability to reason, plan, and adapt to changing circumstances. It is a multidisciplinary field that combines elements of computer science, mathematics, neuroscience, and more.

The Quantum Computing Revolution

The convergence of ML and AI with quantum computing holds immense promise for solving complex problems that are beyond the capabilities of classical computers.

Quantum computers, with their ability to process vast amounts of information simultaneously, can greatly enhance the performance of ML and AI algorithms. They can accelerate training processes, improve pattern recognition, and optimize decision-making.

Quantum machine learning (QML) is an emerging field that explores the potential of combining quantum computing and ML techniques.

QML algorithms leverage quantum properties such as superposition and entanglement to improve the efficiency and accuracy of classical ML tasks.

This synergy has the potential to revolutionize fields such as drug discovery, financial modeling, weather forecasting, and optimization problems.

The Quantum Computing Revolution

Furthermore, AI can be used to enhance the capabilities of quantum computers. Intelligent systems can assist in error correction, optimize quantum circuits, and improve the overall efficiency of quantum algorithms.

AI algorithms can analyze and learn from the behavior of quantum systems, leading to advancements in quantum control and optimization.

As quantum computing continues to evolve, the integration of ML and AI will play a crucial role in unlocking the full potential of this technology.

The synergy between quantum computing, ML, and AI will lead to groundbreaking advancements, enabling us to tackle complex challenges that were previously insurmountable.

In conclusion, the combination of Machine Learning and Artificial Intelligence with quantum computing has the potential to reshape various industries and revolutionize the way we solve problems.

The convergence of these technologies opens up new possibilities, paving the way for a future where intelligent quantum systems can tackle complex challenges with unprecedented speed and accuracy.

Drug Discovery and Material Science

Drug Discovery and Material Science are two dynamic fields that are being revolutionized by quantum computing.

As we delve into the fascinating world of quantum computing, it is important to understand its potential impact on these critical areas of research and development.

In the realm of drug discovery, quantum computing is poised to revolutionize the way we develop and design new medications.

Traditional drug discovery methods are time-consuming and often involve trial and error. However, with the advent of quantum computing, researchers can harness the immense computational power of quantum systems to accelerate the drug discovery process.

The Quantum Computing Revolution

Quantum computers can efficiently simulate the behavior of molecules, allowing scientists to predict their properties and interactions with greater accuracy. This enables the identification of potential drug candidates more efficiently, saving time and resources. Additionally, quantum algorithms can optimize drug molecules, leading to the creation of more effective and targeted therapies.

Material science, on the other hand, deals with the discovery and development of new materials with unique properties. Quantum computing plays a pivotal role in this field by providing powerful tools for simulating and understanding complex materials at the atomic level.

By leveraging quantum algorithms, researchers can perform simulations that were previously impossible with classical computers.

This enables the discovery of novel materials with tailored properties, such as improved conductivity, increased strength, or enhanced catalytic activity. These advancements have far-reaching implications in various industries, including electronics, energy, and healthcare.

The Quantum Computing Revolution

Furthermore, quantum computing can aid in the design of efficient and sustainable materials.

For example, by simulating the behavior of materials under different conditions, scientists can identify materials that are more resistant to corrosion or degradation, leading to the development of longer-lasting and environmentally friendly products.

The integration of quantum computing in drug discovery and material science is not only transformative but also holds immense potential for societal benefits. By accelerating the discovery of new drugs and materials, we can address pressing global challenges such as disease treatment, renewable energy, and environmental sustainability.

In conclusion, the marriage of quantum computing with drug discovery and material science opens up a realm of possibilities. As we continue to unlock the power of quantum computing, we are poised to witness breakthroughs that will shape the future of medicine, technology, and our world as a whole.

Chapter 5: Challenges and Limitations of Quantum Computing

Error Correction and Quantum Decoherence

In the fascinating world of quantum computing, one of the biggest challenges that researchers and engineers face is the issue of error correction and quantum decoherence. While traditional computers are susceptible to errors, the fragile nature of quantum systems makes error correction even more critical. Understanding these concepts is essential for anyone interested in the field of quantum computing.

Quantum decoherence refers to the phenomenon where quantum systems lose their coherence and become entangled with their environment, resulting in the loss of quantum information.

This loss is caused by various factors such as temperature, electromagnetic radiation, and even the slightest vibrations.

The Quantum Computing Revolution

As a result, the delicate quantum state becomes corrupted, leading to errors in calculations and diminishing the computational power of quantum computers.

To combat this challenge, researchers have developed sophisticated error correction techniques. These techniques involve encoding quantum information redundantly in a way that allows errors to be detected and corrected. By introducing redundancy, errors can be identified and rectified without compromising the integrity of the quantum computation.

One of the most promising error correction methods is known as the surface code. This approach uses a two-dimensional lattice structure, where qubits (quantum bits) are placed at the intersections of the lattice.

By measuring the state of neighboring qubits, errors can be detected and corrected. The surface code has shown great potential in mitigating the effects of decoherence, paving the way for more stable and reliable quantum computation.

The Quantum Computing Revolution

While error correction is crucial, it comes at a cost. Additional qubits and computational resources are required to implement error correction codes, making quantum computers more complex and challenging to build. Therefore, finding the right balance between error correction and computational efficiency is an ongoing area of research.

The field of quantum computing holds immense promise for solving complex problems that are beyond the reach of classical computers.

However, the road to realizing this potential is paved with challenges, and error correction is at the forefront of these hurdles. By developing robust error correction techniques and understanding the intricacies of quantum decoherence, scientists are inching closer to unlocking the full power of quantum computers.

In conclusion, error correction and quantum decoherence are critical aspects of quantum computing that must be addressed for the field to reach its full potential. Researchers are working tirelessly to develop innovative error correction techniques such as the surface code, which can help mitigate the effects of decoherence. While challenges remain, the progress made in error correction brings us closer to harnessing the power of quantum computation and revolutionizing various industries.

Scalability and Quantum Gate Operations

In the ever-evolving landscape of quantum computing, one of the most crucial factors is scalability. The ability to scale up quantum systems is essential for achieving larger computations and harnessing the true power of quantum computing. This subchapter delves into the significance of scalability and explores the intricacies of quantum gate operations.

The Quantum Computing Revolution

Scalability plays a pivotal role in quantum computing as it determines the size and complexity of the problems that can be solved. Unlike classical computers that can simply add more transistors to increase computing power, quantum computers rely on the concept of qubits. These qubits are the building blocks of quantum information and can exist in multiple states simultaneously, enabling quantum computers to perform massively parallel computations.

However, qubits are inherently fragile and prone to errors due to environmental interference.

This poses a significant challenge to scaling quantum systems. Researchers and developers are continuously exploring various quantum error correction techniques to mitigate these errors and achieve scalability. These techniques involve redundantly encoding qubits, allowing for error detection and correction.

Another crucial aspect of scalability lies in the optimization of quantum gate operations. Quantum gates are the fundamental building blocks of quantum circuits, similar to logical gates in classical computers.

The Quantum Computing Revolution

They manipulate the quantum states of qubits, enabling the execution of specific quantum algorithms.

Quantum gate operations are highly delicate and must be carefully designed to avoid errors and maintain coherence.

As the number of qubits increases, the complexity of gate operations also amplifies, making scalability even more challenging. Researchers are actively working on developing fault-tolerant quantum gates that are robust against noise and errors.

Furthermore, this subchapter explores the various types of quantum gates, including Pauli gates, Hadamard gates, and CNOT gates, among others.

Each gate serves a specific purpose and contributes to the overall functionality of a quantum circuit. Understanding the characteristics and operations of these gates is crucial for designing efficient quantum algorithms and achieving scalability in quantum systems.

The Quantum Computing Revolution

Ultimately, scalability and quantum gate operations are the cornerstones of the quantum computing revolution. As advancements in technology continue, researchers aim to overcome the challenges of scalability, paving the way for larger and more powerful quantum computers.

By unlocking the potential of scalability and optimizing quantum gate operations, we can harness the true power of quantum computing and revolutionize various fields, including cryptography, optimization problems, and drug discovery, among others.

This subchapter aims to provide a comprehensive understanding of scalability and quantum gate operations, catering to both quantum computing enthusiasts and those new to the field. By grasping the significance of scalability and mastering the intricacies of quantum gate operations, readers can join the journey toward unlocking the power of the future through quantum computing.

Quantum Algorithms and Complexity Theory

The Quantum Computing Revolution

In the world of quantum computing, algorithms and complexity theory play a pivotal role in unlocking the vast potential and power of this groundbreaking technology. This subchapter will delve into the fascinating realm of quantum algorithms and explore how they can revolutionize various fields, from cryptography to optimization problems.

To understand quantum algorithms, let's first grasp the concept of complexity theory. Complexity theory is concerned with the study of how efficiently a problem can be solved. It provides a framework for classifying problems based on their computational difficulty. In the classical computing realm, we have well-known complexity classes such as P, NP, and NP-complete. However, with the advent of quantum computing, a whole new set of possibilities arises.

Quantum algorithms harness the unique properties of quantum mechanics to solve computational problems more efficiently than classical algorithms. One of the most famous examples is Shor's algorithm, which can factor large numbers exponentially faster than the best-known classical algorithms.

The Quantum Computing Revolution

This breakthrough has significant implications for cryptography, as many encryption schemes rely on the difficulty of factoring large numbers.

Another powerful quantum algorithm is Grover's algorithm, which can search an unsorted database quadratically faster than classical algorithms.

This algorithm has far-reaching applications in areas such as data mining, optimization, and machine learning. It promises to revolutionize fields that depend on searching large datasets, enabling us to find optimal solutions in less time.

However, the true power of quantum algorithms lies in their ability to solve problems that are intractable for classical computers. For instance, the quantum approximate optimization algorithm (QAOA) is designed to tackle combinatorial optimization problems, such as the famous Traveling Salesman Problem. QAOA exploits the quantum superposition and entanglement to explore multiple potential solutions simultaneously, offering a promising avenue for addressing complex optimization challenges.

Understanding the complexity of quantum algorithms is crucial for harnessing the full potential of quantum computing. It helps us determine which problems can be efficiently solved using quantum algorithms and which ones are beyond the reach of current technology.

Complexity theory guides the development of new quantum algorithms and helps us classify problems based on their inherent difficulty, paving the way for future advancements in the field.

In summary, quantum algorithms and complexity theory are at the core of the quantum computing revolution. They offer us the tools to solve problems that are otherwise intractable for classical computers and enable us to explore new frontiers in various domains. Whether you are a quantum computing enthusiast, a researcher, or someone curious about the future of computing, delving into the world of quantum algorithms and complexity theory will undoubtedly ignite your imagination and open doors to limitless possibilities.

Chapter 6: Quantum Computing and the Future

Quantum Computing in Industries

As we delve deeper into the realm of quantum computing, the potential applications across various industries become increasingly apparent. The power and capabilities of quantum computing hold the key to revolutionizing sectors such as finance, healthcare, logistics, and more. In this subchapter, we will explore the exciting possibilities of quantum computing in these industries and how it can unlock a brighter future for all.

One of the most promising areas where quantum computing can have a profound impact is finance. With its ability to perform complex calculations at an unprecedented speed, quantum computers can optimize portfolio management, and risk analysis, and even predict market trends with remarkable accuracy.

This technology can potentially revolutionize the stock market, making it more efficient and accessible to investors of all backgrounds.

The Quantum Computing Revolution

In the healthcare industry, quantum computing can contribute significantly to drug discovery and development. The computational power of quantum computers can simulate and analyze complex molecular structures, leading to the discovery of new drugs and treatments for various diseases.

This breakthrough could potentially save countless lives and drastically reduce the time and cost associated with traditional drug development processes.

Logistics and supply chain management are also poised to benefit from quantum computing. The ability to solve complex optimization problems quickly can streamline logistics networks, leading to more efficient routes, reduced fuel consumption, and improved delivery times.

Quantum computing can also enhance inventory management and demand forecasting, minimizing waste and ensuring products are readily available when needed.

Furthermore, quantum computing has the potential to revolutionize cryptography and data security.

The Quantum Computing Revolution

Quantum algorithms can break conventional encryption methods, making it crucial for industries to adapt and develop quantum-resistant encryption techniques.

This technology can safeguard sensitive data, protect intellectual property, and ensure the privacy of individuals in an increasingly interconnected world.

While the potential of quantum computing is immense, it is important to note that we are still in the early stages of its development. The technology is complex, and the challenges that need to be overcome are numerous.

However, as quantum computing enthusiasts, we must embrace this revolution and work towards harnessing its power for the benefit of all.

In conclusion, quantum computing holds tremendous potential for transforming industries across the board. From finance to healthcare, logistics to data security, the advent of quantum computing promises to unlock new frontiers in innovation and efficiency.

As we continue to explore this exciting field, let us envision a future where the power of quantum computing is harnessed to tackle the most pressing challenges of our time and shape a better world for all.

Quantum Computing and Cybersecurity

In this subchapter, we will explore the fascinating intersection of quantum computing and cybersecurity, two rapidly evolving fields that are shaping the future of technology.

As quantum computing continues to advance, it holds the potential to revolutionize various industries, including cybersecurity.

Understanding the implications and challenges in this domain is crucial for both professionals in the quantum computing niche and anyone interested in the broader field of technology.

To comprehend the significance of quantum computing in cybersecurity, let's first delve into the basics of these concepts.

The Quantum Computing Revolution

Quantum computing harnesses the principles of quantum mechanics to perform calculations that are exponentially faster than classical computers.

This computational power could dramatically impact the field of cryptography, which underpins modern cybersecurity. While classical computers rely on mathematical problems that are difficult to solve, quantum computers have the potential to crack these problems with ease.

However, the rise of quantum computing also poses a significant threat to current cybersecurity protocols.

Many encryption algorithms that safeguard our sensitive data today would be rendered obsolete in the face of quantum computing's capabilities.

This necessitates the development of new cryptographic techniques that are resistant to quantum attacks.

The Quantum Computing Revolution

Researchers and professionals in the quantum computing niche are actively exploring quantum-safe encryption methods, such as lattice-based cryptography and quantum key distribution, to ensure data security in the quantum era.

Furthermore, quantum computing can also be leveraged as a tool to enhance cybersecurity. The ability of quantum computers to analyze vast amounts of data and identify patterns could be applied to detect and prevent cyber threats more efficiently. Machine learning algorithms powered by quantum computing could enable faster anomaly detection, real-time threat intelligence, and enhanced predictive analytics, bolstering the overall cybersecurity posture.

However, as we embrace the potential of quantum computing in cybersecurity, we must also address the risks it poses. Quantum computers could potentially break into existing systems and compromise sensitive information. Therefore, it is crucial to develop robust post-quantum cryptographic protocols and ensure a smooth transition to quantum-safe algorithms.

In conclusion, the convergence of quantum computing and cybersecurity presents both opportunities and challenges. Professionals in the quantum computing niche must work alongside cybersecurity experts to develop quantum-safe solutions while ensuring that the power of quantum computing is utilized ethically and responsibly.

For those interested in the broader field of technology, understanding the implications of quantum computing on cybersecurity is essential for staying ahead in the rapidly evolving digital landscape.

Quantum Computing and Machine Learning

In recent years, the field of quantum computing has emerged as a promising avenue for revolutionizing various industries, and one area where its potential is particularly significant is machine learning.

As we delve into the intersection of these two cutting-edge technologies, we witness a realm of possibilities that could unlock the power of the future.

The Quantum Computing Revolution

Machine learning, a subset of artificial intelligence, has already made remarkable strides in transforming our lives, from personalized recommendations on streaming platforms to self-driving cars.

However, traditional machine learning algorithms face limitations when it comes to processing vast amounts of data and solving complex problems. This is where quantum computing comes into play.

Quantum computing harnesses the principles of quantum mechanics to manipulate information in ways that classical computers cannot.

By utilizing quantum bits, or qubits, which can exist in multiple states simultaneously, quantum computers can process and analyze vast amounts of data with unparalleled speed and accuracy.

The potential synergy between quantum computing and machine learning lies in their shared objective of handling and interpreting massive datasets.

The Quantum Computing Revolution

Quantum algorithms specifically designed for machine learning tasks can quickly analyze large-scale datasets, identify patterns, and make predictions more efficiently than classical approaches.

One of the most promising applications of quantum computing in machine learning is in the realm of optimization problems. Many machine learning tasks, such as training neural networks or optimizing complex algorithms, involve finding the best solution among a vast number of possibilities.

Quantum algorithms, such as quantum annealing or quantum support vector machines, have the potential to dramatically accelerate these optimization processes, leading to significant advancements in various fields, including finance, drug discovery, and logistics.

Furthermore, quantum machine learning algorithms have the potential to enhance pattern recognition capabilities, enabling more accurate and robust models.

Quantum-inspired algorithms, such as quantum neural networks, promise to unlock new dimensions of deep learning, allowing for advanced classification, regression, and clustering tasks.

While quantum computing and machine learning hold immense promise, it is important to note that we are still in the early stages of exploring their potential synergy. The technological challenges and limitations of quantum hardware, such as qubit errors and decoherence, need to be addressed for practical implementation.

Nonetheless, the possibilities are tantalizing. Quantum computing and machine learning have the potential to transform industries, solve previously intractable problems, and unlock new frontiers of knowledge. As we continue to unravel the mysteries and capabilities of quantum computing, the future holds exciting possibilities for innovation and progress in the realm of machine learning.

Ethical and Societal Implications of Quantum Computing

The Quantum Computing Revolution

Quantum computing holds immense promise and potential for transforming various aspects of our lives, bringing about a revolution in technology and computation.

However, just like any other groundbreaking technology, it is crucial to understand and consider the ethical and societal implications that come along with it.

As we delve deeper into the realm of quantum computing, it becomes imperative to explore the potential challenges and concerns that may arise.

One of the primary ethical concerns with quantum computing lies in its potential to break traditional encryption algorithms. While this can be advantageous for securing highly sensitive data, it also raises concerns about privacy and national security.

As quantum computers become more powerful, they might be capable of decrypting information that was previously considered secure, leading to potential breaches and unauthorized access to personal or classified data.

The Quantum Computing Revolution

Addressing this ethical dilemma requires the development of quantum-resistant encryption algorithms to ensure data security in the post-quantum era.

Another ethical consideration is the potential impact of quantum computing on employment and labor markets. As this technology advances, it has the potential to automate complex tasks that were previously performed by humans, potentially leading to job displacement or significant shifts in labor markets. It is crucial to anticipate these changes and prepare for the societal implications they may bring. Governments, policymakers, and businesses must collaborate to ensure a smooth transition, providing retraining programs and creating new job opportunities in emerging fields related to quantum computing.

Furthermore, quantum computing raises ethical questions surrounding its use in surveillance and data manipulation. With the ability to process vast amounts of data at unprecedented speeds, quantum computers could enable advanced surveillance systems capable of collecting and analyzing massive amounts of personal information.

The Quantum Computing Revolution

Striking a balance between the benefits of such technologies and the preservation of individual privacy and civil liberties becomes essential.

Additionally, the availability and accessibility of quantum computing technology can create a digital divide. If quantum computers are only accessible to a select few, it could exacerbate existing inequalities and hinder progress for underprivileged communities. Efforts should be made to promote inclusivity and ensure that the benefits of quantum computing are accessible to all, regardless of socioeconomic background.

In conclusion, while quantum computing presents numerous opportunities for scientific breakthroughs and technological advancements, it is crucial to approach its development and implementation with careful consideration of the ethical and societal implications. By addressing concerns related to data security, employment, privacy, and inclusivity, we can harness the transformative power of quantum computing while ensuring a just and equitable future for all.

Chapter 7: Quantum Computing in Practice

Quantum Programming Languages

In the fast-evolving realm of quantum computing, a crucial aspect that holds immense significance is the development of quantum programming languages. These specialized languages have emerged as indispensable tools for harnessing the power of quantum computers and enabling researchers and practitioners to explore the full potential of this revolutionary technology.

Quantum programming languages, much like their classical counterparts, serve as the bridge between human users and quantum hardware. However, due to the fundamental differences between classical and quantum computing, these languages have unique features and capabilities that allow users to leverage the peculiarities of quantum systems.

The Quantum Computing Revolution

One of the key challenges in quantum programming is the concept of superposition, where quantum bits or qubits can exist in multiple states simultaneously. Quantum programming languages provide constructs and operations to manipulate and control these qubits, allowing users to perform calculations and operations that are impossible in classical computing.

Another fundamental aspect of quantum programming languages is the concept of entanglement, which allows qubits to become interconnected in such a way that the state of one qubit is inherently related to the state of another. This phenomenon opens up new avenues for communication and computation, and quantum programming languages provide mechanisms to exploit entanglement for various applications.

Currently, several quantum programming languages have been developed, each with its own unique features and syntax. Some notable examples include Q#, Qiskit, and Cirq. These languages offer a range of high-level abstractions, libraries, and tools that facilitate the development of quantum algorithms and applications.

The Quantum Computing Revolution

While quantum programming languages are still in their nascent stages, they hold immense promise for the future of computing. As quantum technology continues to advance, these languages will play a pivotal role in enabling researchers and developers to unlock the true potential of quantum computers.

For professionals in the field of quantum computing, proficiency in quantum programming languages is becoming increasingly essential.

These languages empower researchers to design and implement quantum algorithms, simulate quantum systems, and optimize quantum circuits.

Moreover, as quantum computers become more accessible, quantum programming languages will enable a wider audience to engage with and contribute to this cutting-edge field.

In conclusion, quantum programming languages are an integral part of the quantum computing revolution.

The Quantum Computing Revolution

They provide the necessary tools and abstractions to harness the power of quantum systems, allowing researchers and practitioners to explore new frontiers in computation and problem-solving.

As the field of quantum computing continues to advance, mastering these languages will be crucial for anyone seeking to unlock the full potential of this transformative technology.

Quantum Software Development Tools

In the fast-evolving field of quantum computing, software development tools play a vital role in enabling researchers and developers to harness the power of this revolutionary technology.

Quantum software development tools provide a set of resources and frameworks that aid in the creation, simulation, and optimization of quantum algorithms, as well as the efficient utilization of quantum hardware.

The Quantum Computing Revolution

Quantum software development tools come in various forms, ranging from software development kits (SDKs) to quantum simulators and programming languages. These tools are designed to simplify the complex nature of quantum computing and make it more accessible to a broader audience, including both quantum experts compute and those new to the field.

One of the fundamental components of quantum software development tools is the quantum simulator. Simulators allow developers to test and debug quantum algorithms without the need for physical quantum hardware.

They provide a virtual environment where users can simulate quantum circuits and observe their behavior. Quantum simulators are invaluable during the early stages of algorithm development, as they enable researchers to iterate quickly and refine their ideas before deploying them on actual quantum computers.

Another essential tool for quantum software development is the quantum programming language.

The Quantum Computing Revolution

These languages are specifically designed to express quantum algorithms and operations, making it easier for developers to write code that can be executed on quantum computers.

Quantum programming languages often have constructs that allow for the manipulation of quantum states, superposition, and entanglement, which are the foundational principles of quantum computing.

Quantum software development kits (SDKs) are comprehensive sets of tools and libraries that provide a unified framework for quantum software development.

SDKs typically include simulators, programming languages, and other utilities that aid in the development and optimization of quantum algorithms. They also provide an interface for interacting with quantum hardware, allowing developers to execute their algorithms on real quantum computers.

While quantum software development tools have made significant strides in recent years, the field is still evolving rapidly.

The Quantum Computing Revolution

Researchers and developers continue to push the boundaries of what is possible with quantum computers, and new tools are constantly being developed to support these advancements.

As quantum computing becomes more mainstream, we can expect to see even more user-friendly and accessible software development tools that will enable a broader audience to unlock the power of quantum computing.

In conclusion, quantum software development tools play a crucial role in harnessing the potential of quantum computing. These tools simplify the complexities of quantum algorithms, provide virtual environments for testing and debugging, and offer programming languages specifically tailored for quantum operations.

As the field of quantum computing continues to advance, it is essential for researchers, developers, and enthusiasts in the quantum compute niche to familiarize themselves with these tools and leverage their capabilities to drive the quantum computing revolution forward.

Quantum Cloud Computing and Quantum Internet

In the ever-evolving landscape of computing, there is a revolutionary paradigm on the horizon - Quantum Cloud Computing and Quantum Internet.

This subchapter aims to introduce the concepts of these cutting-edge technologies and explore their potential implications for the future.

Traditional cloud computing has transformed the way we store, access, and process data. It has enabled unprecedented scalability and flexibility, empowering individuals and businesses alike. However, as we delve into the era of quantum computing, the possibilities for cloud computing expand exponentially.

Quantum Cloud Computing harnesses the power of quantum computers to process complex algorithms and solve problems that are currently intractable for classical computers.

The Quantum Computing Revolution

By leveraging the principles of superposition and entanglement, quantum computers can perform calculations at unparalleled speed and efficiency. This opens up a whole new realm of possibilities, from optimizing supply chains to drug discovery and financial modeling.

Furthermore, the Quantum Internet, a network of interconnected quantum devices, promises to revolutionize information exchange. Traditional Internet protocols rely on classical bits to transmit and process data. In contrast, Quantum Internet employs qubits, the fundamental units of quantum information, to transmit and process quantum states. This enables secure quantum communication, quantum teleportation, and quantum key distribution, paving the way for a new era of secure and efficient data transmission.

The integration of Quantum Cloud Computing and Quantum Internet has the potential to transform industries such as finance, healthcare, logistics, and cybersecurity.

For instance, financial institutions can leverage quantum algorithms to optimize portfolio management and mitigate risks.

The Quantum Computing Revolution

Quantum simulations can revolutionize drug discovery, accelerating the development of life-saving medications.

Quantum cryptography can provide foolproof security, safeguarding sensitive data from malicious attacks.

However, these advancements come with challenges. Quantum computing is still in its nascent stages, and the technology is complex and expensive. Building a robust quantum infrastructure requires substantial investments in research, development, and education. Additionally, ensuring the security and integrity of quantum networks is a paramount concern.

As Quantum Cloud Computing and Quantum Internet continue to evolve, individuals and businesses must stay informed and adapt to this new paradigm.

The potential for innovation and disruption is immense, and those who are proactive in understanding and leveraging these technologies will have a significant advantage in the future.

In conclusion, Quantum Cloud Computing and Quantum Internet represent the next frontier in computing. The fusion of quantum power with cloud infrastructure and networking capabilities holds the key to unlocking unimaginable computational capabilities and secure information exchange.

Embracing and harnessing these technologies will empower individuals and organizations to address complex challenges and unlock the power of the future.

Chapter 8: Quantum Computing and the Quantum Ecosystem

Quantum Startups and Entrepreneurship

In recent years, quantum computing has emerged as one of the most promising fields in technology. Its potential to revolutionize industries ranging from finance to healthcare has attracted the attention of scientists, researchers, and entrepreneurs around the world.

The Quantum Computing Revolution

This subchapter explores the world of quantum startups and entrepreneurship, shedding light on the tremendous opportunities and challenges that lie ahead.

The startup ecosystem has always been at the forefront of technological breakthroughs, and quantum computing is no exception.

Quantum startups are founded by visionary entrepreneurs who recognize the transformative power of quantum technology and are eager to explore its commercial applications.

These pioneers are driven by the desire to harness the immense computational power of quantum computers and develop solutions that can solve some of the most complex problems faced by society today.

However, starting a quantum startup is not for the faint-hearted. The field is still nascent, with many technical challenges yet to be overcome.

The Quantum Computing Revolution

Quantum entrepreneurs face a unique set of obstacles, including the scarcity of quantum experts, the high cost of quantum hardware, and the need for specialized knowledge in quantum algorithms. Nevertheless, the potential rewards for successful quantum startups are staggering, with the possibility of disrupting entire industries and creating new ones.

To navigate this complex landscape, quantum entrepreneurs must possess a combination of technical expertise and business acumen. They must understand the fundamental principles of quantum mechanics while also being able to identify market opportunities and develop viable business models. Collaboration between scientists and entrepreneurs is crucial to bridge the gap between theoretical advancements and practical applications.

This subchapter also explores the emergence of quantum incubators and accelerators, which play a vital role in nurturing and supporting quantum startups.

The Quantum Computing Revolution

These innovation hubs provide access to funding, mentorship, and resources, enabling entrepreneurs to develop their ideas into tangible products and services. They also facilitate collaboration between startups, academia, and industry, fostering an ecosystem where knowledge and expertise can be shared freely.

In conclusion, the world of quantum startups and entrepreneurship is a dynamic and exciting space. It offers immense potential for innovation and disruption, but also presents significant challenges.

With the right combination of technical expertise, business acumen, and support from the quantum community, entrepreneurs have the opportunity to unlock the power of the future and shape the quantum computing revolution. Whether you are a quantum enthusiast, a budding entrepreneur, or simply interested in the future of technology, this subchapter will provide valuable insights into the world of quantum startups and entrepreneurship.

Government Initiatives and Investments in Quantum Computing

In recent years, governments around the world have recognized the immense potential of quantum computing and have taken significant steps to support its development.

Quantum computing, with its ability to solve complex problems exponentially faster than classical computers, has the potential to revolutionize various fields, from drug discovery and optimization problems to cryptography and machine learning. Therefore, governments are investing heavily in this transformative technology to secure their positions in the global race for quantum supremacy.

One of the leading countries in quantum computing research and development is the United States. In 2018, the U.S. Congress passed the National Quantum Initiative Act, which authorized $1.2 billion in funding for quantum research over the next five years.

The Quantum Computing Revolution

This initiative aims to accelerate the development of quantum technologies and maintain U.S. leadership in this emerging field. The funding will support the establishment of Quantum Information Science Centers and the training of a skilled workforce to drive innovation in quantum computing.

Similarly, the European Union has recognized the strategic importance of quantum computing and has launched the Quantum Flagship Program. With an investment of €1 billion over ten years, this program aims to bring together academia, industry, and government to advance European quantum research and development. The program aims to develop a quantum computer with 100 qubits by 2025, positioning Europe as a global leader in quantum computing.

Other countries, such as China and Canada, are also making substantial investments in quantum computing.

China has launched the National Laboratory for Quantum Information Sciences, with a budget of $10 billion, to develop quantum technologies and applications.

The Quantum Computing Revolution

Canada, on the other hand. has established the Quantum Valley Initiative and the Perimeter Institute for Theoretical Physics, fostering collaborations between academia, industry, and government to advance quantum research.

Government initiatives are not limited to funding alone. They also involve policy development and strategic partnerships. Governments are establishing regulatory frameworks and standards to ensure the secure and ethical use of quantum technologies.

They are also forging international collaborations and partnerships to leverage collective expertise and resources in advancing quantum computing.

By investing in quantum computing, governments are not only driving technological advancements but also fostering economic growth and competitiveness. The development of quantum technologies has the potential to create new industries, generate high-skilled jobs, and solve societal challenges that were once deemed unsolvable.

Therefore, it is crucial for governments to continue supporting research, development, and commercialization of quantum computing to unlock the full potential of this revolutionary technology.

In conclusion, governments worldwide have recognized the transformative potential of quantum computing and are investing heavily in its development.

These initiatives aim to accelerate research, foster innovation, and secure global leadership in the quantum computing race. With substantial funding, policy development, and strategic partnerships, governments are paving the way for a quantum computing revolution that will shape the future of various industries and societies at large.

Collaborations and Partnerships in the Quantum Computing Field

In the rapidly evolving field of quantum computing, collaborations and partnerships play a pivotal role in advancing the research, development, and application of this groundbreaking technology.

The Quantum Computing Revolution

The potential of quantum computing to revolutionize industries and solve complex problems has attracted the attention of various stakeholders, including academia, industry leaders, and government agencies. This subchapter explores the significance of collaborations and partnerships in the quantum computing field and sheds light on how these alliances are driving the quantum compute industry forward.

Collaborations between academic institutions, research laboratories, and industry leaders have become essential for the progress of quantum computing.

Quantum research requires interdisciplinary expertise and substantial resources, which often exceed the capabilities of individual organizations.

By pooling their knowledge, expertise, and resources, these collaborations can tackle complex challenges and accelerate the development of quantum computing technologies.

For instance, universities teaming up with technology companies can provide the necessary expertise in quantum physics and algorithms, while industry partners can contribute their expertise in hardware design and optimization.

Partnerships between companies within the quantum computing industry have also become instrumental in advancing the field. Startups are partnering with established technology companies to gain access to resources, funding, and infrastructure needed to bring their innovative ideas to fruition. These partnerships enable startups to leverage the industry experience and market reach of their established counterparts, while larger companies benefit from the fresh ideas and agility of startups.

Moreover, collaborations between quantum computing industry players and government agencies are crucial for the growth of this field. Governments recognize the strategic importance of quantum computing and are investing heavily in research and development initiatives. By working together, industry and government can align their objectives and resources to accelerate the development of quantum computing technologies, foster innovation, and address any ethical, legal, and societal challenges that may arise.

In conclusion, collaborations and partnerships are indispensable in the quantum computing field. They facilitate knowledge sharing, resource pooling, and the acceleration of technological advancements.

Whether it is academia collaborating with industry, startups partnering with established companies, or industry-government alliances, these collaborations are propelling the quantum computing industry forward.

As the quantum computing revolution continues to unfold, fostering collaborations and partnerships will be key to unlocking the full potential of this powerful technology, benefiting all stakeholders and transforming industries across the globe.

Chapter 9: Quantum Computing and the Future of Technology

Quantum Computing and Internet of Things (IoT)

In today's rapidly evolving technological landscape, two cutting-edge fields have emerged as game-changers: quantum computing and the Internet of Things (IoT).

The Quantum Computing Revolution

The convergence of these two transformative technologies holds the promise of revolutionizing our world in ways we could only imagine a few years ago. This subchapter explores the profound impact of quantum computing on the IoT and the limitless possibilities that lie ahead.

The Internet of Things refers to the interconnection of physical devices, vehicles, appliances, and other objects embedded with sensors, software, and network connectivity, enabling them to collect and exchange data.

This interconnected network has already revolutionized various industries, from healthcare and manufacturing to transportation and agriculture. However, the potential of the IoT remains largely untapped due to the limitations of classical computing.

Enter quantum computing, a paradigm-shifting approach that leverages the principles of quantum mechanics to perform complex computations at an unprecedented speed. Quantum computers have the potential to solve problems that are currently intractable for classical computers, making them an ideal tool to unlock the full potential of the IoT.

The Quantum Computing Revolution

One of the most significant challenges in the IoT is the sheer volume of data generated by billions of interconnected devices. Classical computers struggle to process and analyze this staggering amount of information efficiently.

Quantum computing, with its ability to process vast amounts of data simultaneously, can provide the necessary computational power to handle this data deluge effectively.

Moreover, quantum computing can enhance the security of the IoT ecosystem. As the number of connected devices increases, so does the vulnerability to cyber threats. Quantum computers have the potential to break traditional encryption algorithms that secure our data. However, they can also offer quantum-resistant encryption methods that are virtually unhackable, ensuring the privacy and security of IoT networks.

Furthermore, quantum computing can optimize the efficiency of IoT systems. By leveraging quantum algorithms, complex optimization problems, such as route planning, resource allocation, and energy management, can be solved in real-time, leading to increased efficiency and cost savings.

The combination of quantum computing and IoT opens up endless possibilities for innovation and disruption across industries. Imagine smart cities with optimized traffic flow, energy grids with real-time demand management, and healthcare systems with personalized treatments based on real-time patient data analysis. These scenarios once considered a distant dream, can become a reality through the convergence of quantum computing and IoT.

As the quantum computing revolution gains momentum, individuals and organizations in the quantum computing niche need to grasp the potential impact of this convergence. By understanding the synergies between quantum computing and IoT, we can harness their power to shape a future that is smarter, more connected, and more efficient than ever before.

Quantum Computing and Big Data Analytics

In this subchapter, we delve into the intriguing intersection of quantum computing and big data analytics, exploring the immense potential and promising future that lies ahead.

The Quantum Computing Revolution

As we embark on the quantum computing revolution, it is crucial to understand how this groundbreaking technology can revolutionize the way we analyze and extract insights from massive datasets.

Big data analytics has become an indispensable tool for businesses, scientists, and governments across various sectors. However, as the volume and complexity of data continue to grow exponentially, traditional computing methods struggle to keep up with the demands of processing and analyzing such vast amounts of information.

This is where quantum computing steps in, offering a game-changing solution to tackle the challenges of big data analytics.

Quantum computers harness the power of quantum mechanics, allowing them to perform computations at an unprecedented scale and speed. Unlike classical computers that process information in binary bits (0s and 1s), quantum computers utilize quantum bits or qubits, which can exist in multiple states simultaneously, thanks to the principle of superposition.

The Quantum Computing Revolution

This inherent ability of qubits to hold and process multiple pieces of information simultaneously gives quantum computers a tremendous advantage in handling complex datasets.

With quantum computing, big data analytics can be revolutionized in several ways. Firstly, quantum algorithms can efficiently search and analyze vast datasets, enabling rapid identification of patterns, trends, and correlations that would be otherwise hidden or time-consuming to uncover. This capability has immense implications for various industries, including finance, healthcare, and cybersecurity, where data-driven insights are crucial for decision-making and risk assessment.

Moreover, quantum computing can enhance machine learning algorithms, enabling faster and more accurate predictions and recommendations. By leveraging quantum algorithms, data scientists can train complex models on large datasets more effectively, leading to improved accuracy and efficiency in tasks such as image recognition, natural language processing, and personalized recommendations.

The integration of quantum computing with big data analytics also opens up new possibilities in data privacy and security. Quantum cryptography algorithms can provide a higher level of encryption, ensuring sensitive data remains secure from potential attacks. Furthermore, quantum machine learning algorithms can help identify anomalies and potential security breaches in real-time, enhancing the overall cybersecurity framework.

In conclusion, the convergence of quantum computing and big data analytics presents a transformative opportunity for businesses and researchers alike.

By leveraging the power of quantum computers, we can unlock the full potential of big data, revolutionizing the way we analyze, interpret, and utilize vast amounts of information. The implications are far-reaching, promising advancements in various domains and ultimately shaping the future of technology and innovation.

Quantum Computing and Blockchain Technology

The Quantum Computing Revolution

In recent years, both quantum computing and blockchain technology have emerged as groundbreaking fields with the potential to revolutionize various industries. While they may seem unrelated at first glance, the convergence of these two technologies holds immense promise for the future. This subchapter aims to explore the intersection of quantum computing and blockchain technology, shedding light on their potential applications and the impact they can have on the world.

Quantum computing, with its ability to process vast amounts of data simultaneously and solve complex mathematical algorithms, has the potential to enhance the security and efficiency of blockchain networks. Blockchain technology, on the other hand, offers transparency, immutability, and decentralization, making it an ideal platform for secure and trustworthy transactions. By combining the strengths of both technologies, we can unlock a new era of advanced computing and secure data storage.

The Quantum Computing Revolution

One of the most significant challenges faced by traditional blockchain networks is the computational power required for complex mathematical calculations, such as those used in cryptographic algorithms. Quantum computers can perform these calculations exponentially faster than classical computers, potentially rendering current cryptographic methods obsolete. However, this also presents an opportunity to develop quantum-resistant cryptographic algorithms, ensuring the security of blockchain networks in the quantum era.

Moreover, quantum computing can enable more efficient consensus mechanisms in blockchain networks. Currently, popular consensus algorithms like Proof of Work (PoW) and Proof of Stake (PoS) consume significant amounts of computational resources and energy. Quantum computing could introduce new consensus models that are faster, more secure, and more environmentally friendly, addressing the scalability and sustainability issues faced by blockchain networks today.

The Quantum Computing Revolution

Additionally, quantum computing can enhance privacy in blockchain applications. By leveraging quantum entanglement and superposition, quantum-resistant encryption methods can be developed, protecting sensitive data from unauthorized access. This can be particularly transformative in industries such as healthcare and finance where data privacy is paramount.

Furthermore, the synergy between quantum computing and blockchain technology can enable the creation of more sophisticated smart contracts.

These self-executing contracts, built on blockchain platforms, can be enhanced by quantum computing's ability to process complex logic and optimize decision-making processes. This opens up possibilities for advanced applications in supply chain management, decentralized finance, and digital identity verification.

In conclusion, the convergence of quantum computing and blockchain technology offers a transformative path for the future. By harnessing the power of quantum computing, blockchain networks can become more secure, scalable, and privacy-preserving. The combination of these two cutting-edge technologies has the potential to disrupt various industries and unlock new opportunities for innovation. As we delve deeper into the quantum computing revolution, understanding the intersection with blockchain technology becomes crucial for all individuals and organizations involved in the quantum computing niche.

Chapter 10: Conclusion and the Quantum Computing Revolution

Summary of Key Concepts

The Quantum Computing Revolution

In this subchapter, we will summarize the key concepts explored throughout the book "The Quantum Computing Revolution: Unlocking the Power of the Future." This summary is designed to provide a comprehensive overview of the fundamental ideas and principles related to quantum computing, making it accessible to readers of all backgrounds, while also catering to the specific interests of the quantum computing niche.

Quantum computing is a rapidly evolving field that promises to revolutionize information processing, offering unprecedented computational power beyond the capabilities of classical computers. To understand the potential of quantum computing, it is essential to grasp several key concepts.

Firstly, we delve into the principles of quantum mechanics, the branch of physics that underpins quantum computing. Quantum mechanics challenges our classical intuition by introducing concepts such as superposition and entanglement. Superposition allows quantum bits or qubits to exist in multiple states simultaneously, exponentially increasing computational possibilities.

The Quantum Computing Revolution

Entanglement, on the other hand, creates correlations between qubits, enabling the manipulation of information at a quantum level.

Building upon these principles, we explore quantum gates, which are analogous to the logic gates in classical computing. Quantum gates manipulate the state of qubits, allowing for the execution of quantum algorithms. We discuss essential gates like the Hadamard gate, CNOT gate, and the Pauli gates, emphasizing their significance in quantum computations.

Next, we delve into quantum algorithms, highlighting their potential to solve complex problems more efficiently than classical algorithms.

We provide an overview of prominent algorithms such as Shor's algorithm for factoring large numbers and Grover's algorithm for searching databases, showcasing their impact on cryptography and optimization.

The Quantum Computing Revolution

To address practical concerns, we explore the current state of quantum hardware and the challenges faced in scaling up quantum computers. We discuss different types of qubits, including superconducting, trapped ion, and topological qubits, and their respective advantages and limitations. Moreover, we analyze the concept of quantum error correction, a crucial aspect for maintaining the integrity of quantum information.

Lastly, we touch upon quantum applications in various fields, including finance, healthcare, and machine learning.

We discuss how quantum computing can optimize financial portfolios, accelerate drug discovery, and enhance artificial intelligence algorithms.

In conclusion, this subchapter provides a concise summary of the key concepts covered in "The Quantum Computing Revolution:

Unlocking the Power of the Future." Whether you are a quantum computing enthusiast or a newcomer to the subject, this summary serves as a valuable reference, offering a solid foundation for exploring the vast potential of quantum computing.

Impacts of Quantum Computing on Society

Quantum computing is a revolutionary technology that has the potential to transform various aspects of our society. It promises to unlock the power of the future, enabling us to solve complex problems that are currently beyond the reach of classical computers. The impacts of quantum computing on society are vast and far-reaching, and they have the potential to reshape industries, accelerate scientific discoveries, and revolutionize the way we live.

One of the key areas where quantum computing will have a significant impact is the field of cryptography. Quantum computers can break many of the encryption algorithms that are currently used to secure our sensitive data. This poses a significant risk to our digital infrastructure, including financial transactions, communication networks, and even national security. As quantum computers become more powerful, we will need to develop new encryption methods that are resistant to quantum attacks.

The Quantum Computing Revolution

Another area that will be transformed by quantum computing is drug discovery and materials science. Quantum computers can simulate complex chemical reactions and molecular structures with unprecedented accuracy. This will greatly accelerate the process of drug discovery, allowing researchers to identify potential candidates more quickly and accurately. Similarly, quantum computing can help optimize the design of new materials with specific properties, revolutionizing industries such as energy, electronics, and transportation.

The impact of quantum computing on artificial intelligence (AI) is also profound. Quantum machine learning algorithms have the potential to outperform classical machine learning methods, enabling us to develop more advanced AI systems. Quantum computers can process and analyze large datasets much faster, leading to more accurate predictions and insights. This will have implications for various industries, including finance, healthcare, and autonomous systems.

Moreover, quantum computing will play a crucial role in addressing some of the world's most pressing challenges, such as climate change, optimization, and logistics. Quantum algorithms can help optimize complex systems, such as supply chains and transportation networks, leading to significant efficiency improvements. In the field of climate modeling, quantum computers can help simulate the behavior of complex systems, allowing us to better understand and mitigate the impact of climate change.

In conclusion, the impacts of quantum computing on society are vast and transformative. From revolutionizing cryptography to accelerating scientific discoveries and optimizing complex systems, quantum computing has the potential to reshape industries and improve our lives in countless ways. As this technology continues to advance, society must prepare for the challenges and opportunities that lie ahead.

The Path Forward: Embracing the Quantum Computing Revolution

The Quantum Computing Revolution

Welcome to the future! In this subchapter, we will explore the exciting and transformative world of quantum computing and how it will revolutionize our lives. Whether you are a quantum computing enthusiast or someone just curious about the potential of this groundbreaking technology, this chapter is for you.

Quantum computing represents a paradigm shift in our approach to computation. Unlike classical computers, which use bits to process information, quantum computers utilize quantum bits or qubits.

These qubits can exist in multiple states simultaneously, allowing for parallel processing and exponentially increasing computational power.

The potential applications of quantum computing are vast and varied. From solving complex optimization problems to simulating chemical reactions, from enhancing machine learning algorithms to revolutionizing cryptography, the possibilities are seemingly endless. Industries such as finance, healthcare, logistics, and even entertainment stand to benefit greatly from the power of quantum computing.

The Quantum Computing Revolution

However, the path forward is not without challenges. Quantum computing is still in its early stages, and scientists and researchers are actively working on resolving key technical hurdles.

These include improving qubit stability, reducing decoherence, and developing fault-tolerant quantum systems. Additionally, there is a need to build a robust ecosystem around quantum computing, including developing programming languages, tools, and algorithms specifically designed for quantum systems.

To fully embrace the quantum computing revolution, it is essential to foster collaboration between academia, industry, and government agencies. This collaboration will enable the sharing of knowledge, resources, and expertise, facilitating the rapid advancement and adoption of quantum computing technologies.

For individuals interested in quantum computing, this is an opportune time to get involved.

The Quantum Computing Revolution

Learning the fundamentals of quantum mechanics, familiarizing oneself with quantum algorithms, and gaining hands-on experience with quantum computing platforms will be invaluable in this emerging field.

There are numerous online resources, courses, and communities available for anyone interested in exploring and understanding quantum computing.

In conclusion, the quantum computing revolution is upon us, and the path forward is filled with immense potential. It is an exciting time to be part of this transformative journey.

By embracing this revolution, we can unlock the power of the future and pave the way for a new era of innovation and discovery. So, let us embark on this adventure and witness the incredible possibilities that quantum computing has to offer.

Quantum Compute

Once we introduce Quantum computing life will change for everyone. The government will transform into an immense power like we have never seen before.